[Astronomical Energy Resources, Value, and Lifestyles]

(Some collected satires, essays, and papers)

by

HORACE W. CROSBY, JR., M.Ed.

Engineering Unlimited Energy Resources and Value

The Curing Of Diseases

Economic Growth And Longevity

Come; Ode To Real Doing

The Idea Of A 'Slap Trance' In Clinical Hypnotherapy

Energy Supplies Obsessions

The Benefits Of Controlling Other Governments

Democratizing The Election Process

T.V. Votin'

The Images Of Our Presence In Some Nato Countries Today

Some Kinds Of Social Programming In An Age Of Reason
And Improvement

An Economic Lament

Regular Periods

About the Author

ISBN: 146790290X
ISBN 13: 9781467902908

Dedication

To the United States Veterans of Wars and their Spouses, who did so much to add to and improve our systems of business organizations, information dissemination, communications, health, and military security throughout the World in this century.

Horace W. Crosby, Jr.

Author's Note

I hope these papers and satires about Astronomical Energy Resources, Value, and Lifestyles are funny. Some friends of mine all laughed at each paper; if not outside belly laughs, at least inside chuckles.

Each paper can be read separately and each one doesn't take too long to read.

Horace W. Crosby, Jr.

Foreword

As a Personality and Counseling Theories and Therapies expert, it occurred to me that a 'good mood' and a good quality of life are what it is all about. Things like goods, services, health, long life, opportunities, energy, money, and free wills seemed to be related to that.

As a good citizen, I decided that the best way I could contribute academically, and apply some of my theories, would be to write some things that would positively affect more good moods and make better qualities of life available for good folks in general.

How could I contribute as an academician in a scholarly-like way, translate academic information into common terminology, and be funny at the same time?

First I thought the major problem was the money supply, then funding for medical science, and then the energy supplies. It's probably all three, and more.

I've never enjoyed economics that much, and I'm not good at microeconomics, which seems to be holding us back.

The solutions for the World's energy needs and wants are more related to macroeconomics; like using astronomical sized mechanisms to gather, store, transform, use, sell, revitalize, save, and/or recycle the daily powers that come to us from the Sun, the Moon, and the spinning of the Earth, to separate Ocean water into fuels like hydrogen gas and oxygen gas. The Oceans, then, can be seen as a gigantic fuel tank. The problem is to get the energy to access the fuel.

These forces that come to us are literally astronomical, so the machines we're going to have to build are going to have to be gigantic.

Engineering Unlimited Energy Resources and Value

Envision taking the biggest pyramid in Egypt and turning it upside down. Get enough of those pyramids to create an enormous sphere, with the bottoms of the pyramids on the outside of the sphere, and all the tips focused towards the center.

Put half of the sphere underwater at sea level, and the top above sea level. It would look like a giant dome above the ocean.

Put mirrors on the bottom side which is underwater, so the Sun will reflect back to the thermal collector at the center, significantly increasing the amount of heat focused, magnified, and collected.

The Sun will focus, from sunrise to sunset, through the magnifying glasses on the dome, and will reflect off the mirrors on the underside dome. These two powerful mechanisms would focus a tremendous amount of heat energy on the thermal collector at the center. This heat would be a lot hotter than magma from a volcano. If it was not, according to simulated computer models then increase the size of the sphere.

Use the heat from the thermal collector to change the Ocean water into pure water and some other byproducts, using two available processes. The division of water into hydrogen gas and oxygen gas can be accomplished using the process of electrolysis. The electrical power necessary to run the process of electrolysis can be created using the process of photovoltaics, using the heat from the sun to create electricity. There would be plenty of heat to vaporize the pure water into hydrogen and oxygen gases, and to make products from different kinds of left over residue, and to recycle any waste. Also, photolysis could use thermal energy to convert pure water into hydrogen and oxygen ions. Then, hydrogen gas could be produced.

Engineering Unlimited Energy Resources and Value

Using formulas and methods in Physics and Chemistry, there would be 'no end' to what we could produce, and some Chemists and/or Physicists can show somebody who does not believe this is true, how to prove it for oneself.

If we needed to, we could simply use the thermal energy to pump ocean water through big pipes on the Ocean floor, to a dome on nearby land, where the ocean water could be separated into different products, especially for safety reasons.

It sure would help out with the problem of providing fuel for the World, and produce some valuable products to sell, if this kind of light magnification resource was engineered.

The Curing
of Diseases

Let's assume that we went ahead somehow and built the gigantic mechanisms to gather that much energy having unbelievable monetary value. There would be so much that we could do some important things for good folks in general.

What about trying to cure diseases?

The problem of curing diseases touches on one of the most profound considerations of us all: the ability to reduce the power of those phenomena which threaten human life and health.

Over the years, from time to time, some people have found 'wonder drugs', and other treatments which help us fight diseases. More progress has been made toward resolving the problems of diseases within the last hundred years in the evolution of humanity, than in all previous years combined!

Human knowledge is expanding exponentially in many areas, partly due to the impact of our inventions and systems of education, business, and communication.

Medical science has developed a basic process for dealing with diseases; using a historical, interdisciplinary, scientifically oriented approach to finding the most painless, safest, fastest cure for every disease known to us.

The ideal healing process is simple: (1) find the disease, (2) find the phenomena which produce the disease, (3) find those phenomena which will keep the disease from diminishing life and health, and (4) administer those chemicals, symbiotic organisms, light waves, sound waves, etcetera, which will get the job done, with minimum trauma to the person.

More and more people these days are learning about DNA (deoxyribonucleic acid), a complex molecule which exists in all living cells. DNA carries the codes which 'tell' the cell exactly which atoms and molecules to combine in what order and under what conditions.

The Curing
of Diseases

In each cell of a human body, there is a DNA compound which holds the key to the systematic maintenance and reproduction of every part of the body. So it's easy to see why scientists are constantly looking for something that will affect the DNA of disease cells in our bodies and, at the same time, not hurt the DNA in our healthy cells. The process is difficult, and 'hit and miss' at best sometimes.

Years ago, knowledge of the effect of DNA on organisms' life was unavailable, and there were no histologists, microbiologists, geneticists, electron microscopes, magnetic resonance machines, or computers.

Physicians, nurses, and some others have had an almost impossible task with not enough resources.

Things are different now. The human genome, for example, has been mapped. Death 'fears' us more and more.

Medicine has failed us, as I see it, by not achieving the degree of interdisciplinary cooperation necessary to solve some of today's complex medical problems. Also, too often, medical efforts have been directed towards looking for 'the' drug that will work, like the polio vaccine works against polio, rather than a treatment or a combination of medications.

In the long run, however; each disease that attacks a person's body is going to have to be cured taking into consideration the uniqueness of the disease, the uniqueness of the person, and the external environment.

Say, for instance, that someone has "cancer" ---- a unique cancer ---- a DNA structure unlike any other ever experienced. What would one do?

One has many courses of action, from euthanasia, to prayer, and more.

What if a physician could extract, or take a picture of some cancer cells from the person, and do the same to some cells from the part of the person's body which the cancer was invading, and give these to a molecular geneticist?

The Curing
of Diseases

The Geneticist would discover the similarities and differences between the two types of cells. To help, a centrally located library of molecular disease pictures, types of DNA attacked by them, successful treatments, and prognoses should be available. This could help determine which treatments would alter or destroy the DNA of the invader, and not the DNA of the person.

With the present state of our technology, medical progress for this type of process is being researched, tested, refined, and administered today. If only they had more money (energy which represents monetary value to businesses willing to invest).

Could a process like this be used to find those treatments which would facilitate human development by keeping someone's DNA from being harmed? (much like a 'cell callous') Could it enhance the possibilities for its maintenance and development, giving new abilities to old cells? We face many ethical considerations in DNA technology.

What are the barriers? What are the things we can do to develop knowledge and skills in the areas necessary to resolve the problem of curing diseases? Cost and time are the major considerations.

Can you imagine how much it would cost in terms of money and energy to hire a molecular geneticist/researcher/physician to use the lab equipment, publicly or professionally accessible information resources, and computer time necessary to solve each person-disease problem?

Put a dime in the 'healing booth'. (Could it have something to do with large mechanisms to gather, store, transform, use, sell, revitalize, save, and/or recycle those energies and related products; and funding for medical science?).

Economic Growth and Longevity

(a satire)

So, let's assume again, that businesses, for example, changed the meaning of 'value' to include the unlimited value of the resources we have from the powers of the Sun, the Moon, and the spinning around of the Earth at 1,000 miles per hour.

What would it take to convince us to live longer?

According to my calculations, we need to randomly select .068% of our population, and give non-communicable deadly disease to our society over a four week period. Assuming an average life expectancy of 70 years and an average age of death of 40 years when we introduced the disease, at least 400,000,000 years of life would be lost in one month, considering the size of our population here in the United States of America at the time of these writings.

Random selection would insure fairness to all, and decrease bickering.

The technique would be easily accepted, since selective genocide is and has always been so commonplace. The benefits, however; would have to be explained (although not necessarily) to the public.

The impact of such a sudden loss would have a profound effect on the remaining populace. We would be in a state of shock. The spirit of national cohesiveness would increase significantly. Medical research would probably receive a lot of money really fast.

Due to the dramatic increase in the number of coffins, the funeral business would be a good short term investment.

Of course, the clergy would have to put in a lot of overtime.

So, as I figure it, as it is now, if the current economic growth rate is about 3% per year; given the state of American technology and the rate of advancement of medical science; within 30 years, we are going to increase the average life expectancy to around 130 years of age.

Economic Growth and Longevity

If we had the money to give to the medical community, maybe it could be done in 10 years, and we could add some years of life for ourselves.

So, assuming an average life expectancy of 70 years for now, if we extended it to 130, we could add 60 years to each person's life. If we did this 20 years earlier than we can now; in thirty years we will have added about 24,009,000,000 years of life. (Maybe we could borrow the money to do it and raise the deficit.)

This would be a net gain of around 23 billion years of age as a group, considering the initial 400 million lost. Not bad.

But then we'd have to worry about overpopulation and pass laws to protect ourselves.

Now, let's look at it another way. With all that free unlimited energy as a form of business value and all the additional products and services produced, it just might be possible to increase the present rate of GNP to 9% per year instead of 3%. This would also increase the amount of money to invest in medical research so we could live to be 130 years of age on the average.

Why not increase it 90% in one year? Who knows? Do the medical scientists think they could increase the standard life expectancy to 130 years with adequate funding in one year?

On the other hand, 23 billion years more or less, are insignificant, anyway, given our present quality of life. Let's keep things the way they are. If it works, don't fix it. Besides, according to some current microeconomic models, we've only got so much money to go around, and we're running out of resources, and the things we do have and own are getting less and less valuable as time goes on.

(How many years, how fast, can we add to each person's life if we have the money to spend on it?)

Come; Ode to Real Doing

Are these visionary ideas really 'doing anything' for anybody so far? I think it's time to lighten up a little and consider whether or not I've made some 'good sense'.

Come; Ode to Real Doing

(a-poetic apology)

I think, I know, I stink,

therefore I think I know I am;

and so is, too, sometimes,

to a degree; real doing. And what

would all the forms of real doing be?

Thinking, knowing, stinking:

those are only three.

Do be

do be

do be do.

INTENDED MEANINGS

The poet is: 1. asking all the meanings of the term 'real doing' to come to him; 2. attempting to offend one's sensibilities to a degree, to get the attention of the

Come; Ode to Real Doing

reader; 3. requesting that the reader analyze the meanings of the word 'so', as it relates to real doing; 4. asserting that he is, as well as each of the meanings of the word 'so', sometimes, to a degree, a form of real doing; 5. apologizing for not being able to cover the subject of real doing; and 6. apologizing for not really doing anything except writing a poem.

Symbolically, 'come' is emphatically telling 'so' that 'come' knows 'so' stinks, too, to a degree, however; there is more to real doing than stinking, knowing, and thinking, and 'come' is apologizing for not knowing. All 'come' can do is write an ode, and not a very good one at that. 'Come', in terms of punctuation, is telling 'so' to pause. The word 'too' is also symbolically represented. Is 'so' saying 'forever', 'everything', 'now'? 'So' spits 'too', too. Is 'come' saying 'how fast, let's have fun'? Is one 'spitting' and the other 'blanking'? Could this be a way of having fun? Is fun a form of real doing?

Could this be a 'dig' to some philosophers and existentialists?

Is the writer asserting that he, too, is a poet; a real doing poet? Does the writer seem to hate the terms 'sometimes' and 'to a degree', and is he sick and tired of being told he isn't 'really doing anything', on this road to free unlimited energy resources?

Is he saying, in essence '_____' real doing?

(NOTE: The poet would consider it slanderous to publicly call this poem 'Ode to the Commode', or something like that.)

The Idea of a 'Slap Trance' in Clinical Hypnotherapy

(a spoof)

It occurred to me that the only thing that might convince some people to build the kinds of enormous mechanisms we need, so that we will have access to the amount of energy it is going to take to 'run' many of the systems of the World, is to hypnotize them.

As Milton Erickson says, "It's easier to put a subject in a particular kind of trance as therapy progresses." (or something like that)

Is a trance state something like a state of day dreaming, age regression, meditation, dreaming, catalepsy, anesthesia, role playing, goal-directed imagining, communication, amnesia, social programming, hallucination, dissociation, ideo-motor functioning, or post-hypnotic suggestion(s)?

As I understand it, the hypnotist first guides the conscious and/or unconscious attention(s) of the subject to the hypnotist's desired area(s) of focus ... .

Then; when the desired state of trance is achieved, the hypnotist suggests something to the subject. (Sometimes you have to pause or slow down to give a suggestion time to sink in ... .)

If the subject later experiences and/or expresses unconsciously and/or consciously the suggestions of the hypnotist, the hypnotism works. If not, it doesn't. Or maybe it might work sometimes to a degree.

The type of trance 'opens' or 'closes' the subject's will to a degree to the suggestion(s) of the hypnotist.

The 'Slap Trance' may be a useful technique in clinical hypnosis, especially for use in hypnotizing a subject who is less powerful; and when the hypnotist wants the subject not to experience and/or express something.

The Idea of a 'Slap Trance' in Clinical Hypnotherapy

This may not be an effective technique when used for someone who is more powerful, or who is inclined to disagree with, or to do the opposite of what is suggested, especially when suggested by an authority. (the 'will not to')

Take the example of a parent and a child who will not behave.

According to my observations and study of clinical hypnosis; at this point the parent ... , being the more powerful of the two, may choose to use the Slap Trance technique. (Eliciting surprise sometimes helps, or yelling at the child, or 'looking down' to her or him, or staring hard at the child, or lowering the tone of the parent's voice, or speaking slowly with volume.)

Then, the parent might suggest to the child that the child not experience and/or express in a certain way, in a particular psychologically defined behavioral setting, and under a given set of circumstances, for example.

In addition to that, the parent might suggest an alternative form of experiencing and/or expressing.

If the child later experiences and/or expresses the suggestion or suggestions of the parent; the Slap Trance may have been effective to a degree in achieving the induction of the hypnotic suggestion(s).

(Sometimes, a sense of humor, or a reference to an authority, or a degree of repetition or redundancy is necessary, as well as the use of the child's own modes of expression, and direct or implicit focusing on agreements and similarities between the child and the parent.)

If hypnosis is not achieved; perhaps the therapy should be continued. Of course, according to Milton Erickson, as I understand him, if the trance is required again; the technique may be used 'as if' the parent is intending to entrance the child. The child may then immediately go into the trance and be open to the suggestion(s) of the parent.

The Idea of a 'Slap Trance' in Clinical Hypnotherapy

In a psychologically defined behavioral setting, how many different kinds or combinations of hypnotic trances are there; and what is the frequency, duration, and intensity of each during some intervals of time within the period of a year; with respect to eliciting what kinds or combinations of experiences and/or expressions with what concomitant and future effects for what and whom?

Maybe the conscious sense of identity is hypnotized by daily repetitious trance-like stimuli, and the attention of the sense of self needs to be, sometimes, to a degree, dehypnotized into a hypnotic unconscious trance, resulting in a state of rest or sleep.

Can hypnosis be used as a method of unlearning harmful habits, improving concentration, improving memory, discovering neuroses, learning, teaching, establishing and maintaining helpful habits and feelings, for examples?

Can it be used for improving problem solving abilities, releasing creative processes, escaping harmful social restraints, or coping with phobias, or compulsions?

Could it be used to help a person deal with obsessions, increase personal motivation, forget painful memories for a time, decrease worries, expitate painful memories and associated experiences and expressions, escape the problems of the world for a while, find the good mood part of oneself and experience and/or express that for an extended period of time, or to just have fun, or to relieve many aforementioned additional kinds of psychological afflictions?

Who wants to do the will someone else? Who wants to be with someone who may make you afraid because of the powers they have? What if the hypnotist misinforms you, or tries to manipulate you? Also, painful material might surface and stay at the forefront of one's conscious mind.

Maybe some types of hypnotic transactions should be resisted and harmful transference should be sometimes expressed to the hypnotist. One should be

The Idea of a 'Slap Trance' in Clinical Hypnotherapy

reminded though ... that spontaneous harmful reverse conscious and/or unconscious transactional transference afflicting the persona of the subject may occur.

The Slap Trance is only one kind of trance. There is more than one kind of Slap Trance. Are some helpful and/or harmful? (such as deliberate 'wind-breaking', or unwinding, for instance)

Maybe evidence of the effects and fairness of the Slap Trance should be observed and measured; in a precise, controllable, predictable, repeatable way. It could be a useful hypnotic tool.

Could the Slap Trance be used to convince some that there is truly a real road to free unlimited resources, and that these powers are here and available for use?

Energy Supply
Obsessions

What if one's mind kept repeating the sentence "Oh God, I'm surrounded by microeconomic money vampire clones." redundantly? Historically, desires for vast amounts of energies in one form or another have been some of the most important factors in some major governmental and business decisions; resulting in things like wars and major depressions. Energy supply obsessions can be very, very serious, and evidently, even life-threatening sometimes.

Energy supply obsessions; definitely a combination of neurotic afflictions, may be related to a degree to having to deal with a personality type best described by a place near the anal section of the body. Symptoms vary and may manifest themselves in combinations of ways.

If one expresses the pain of the people of the World, and their anger, impatience, anxiety, and other manifestations of severe disorders related to the lack of sufficient energy resources, one might be manifesting some of the symptoms of energy supplies obsessions.

The symptoms might manifest themselves as an absence of good public sense when it comes to voting about the construction of large scale mechanisms to gather, store, transform, use, sell, revitalize, save, and/or recycle energies from the Sun, the Moon, or the spinning around of the Earth, for examples.

Keep in mind that these supplies are free and unlimited, and that our future generations will still have them available, after we are dead and gone.

It is a serious problem for many current Personality Theorists and Therapists to explore the causes, symptoms, and treatments for each kind or combination of psychologically defined energy supply obsessive disorders.

Maybe some people are 'hypnotized' by daily harmful repetitious trance-like stimuli related in part to a money neurosis and impatience, because of living in a microeconomic 'scam'; and resulting in a person working, except not being able to buy and own land, a home, and other material things in return.

Energy Supply
Obsessions

I don't know. It might help to keep 'in spite of' all this, let's have fun anyway, be comfortable, relax, (and not the opposites), at the forefront of one's mind, and sometimes do your favorite things for a while.

The Benefits of Controlling Other Governments

(a scathingly sarcastic paper)

Some other countries have historically and recently hurt some of our citizens and offended the sensibilities of others.

Take the case of Iran, for instance, at the time of these writings.

Personally, I am sick and tired of us being called names like 'The Great Satan'. Maybe we should start calling *them* names.

I don't think we should bomb them though, because so many of them are comparatively not harmful, and are good folks, just like us.

Is it our fault that their leadership are somewhere between lunatics and raving maniacs?

Is it our fault that the wealth of the country does not trickle down to the populace, and that they, in general, are miserable, compared? Did we force them to go to war against another country? Do we teach our children that it is ok to be unethical as long as it's legal? Do we condone immoral business practices because that is what was done to us or because that is the way the 'real world' is? Wait a minute ... that sounds kind of familiar to me, as if I've been through something like this before, right here in the good old USA.

Oh well, anyway, maybe this would work, in a country like Iran, or Syria: Get together with some NATO forces and:

1. select those individuals who are most responsible for destabilizing democratic forces;

2. annihilate the existence of those individuals, or something like that;

3. help pro-democratic Leaders to re-organize their government as a country guided by democratic principles;

The Benefits of Controlling Other Governments

4. take whatever resources we deserve from the country for helping the good folks who were suffering;

5. give control of the government back to politically qualified governmental administrative Leaders;

6. leave them hopefully in peace; and

7. trade with them and invest money in their businesses and businesspeople, so we, and the people there, can have better standards of living.

British, Spanish, French, and Soviet colonialism are precedents for these types of actions.

What would be some of the benefits of these kinds of actions?

Maybe less people would want to emigrate here. Maybe we could make some money by using their labor. Maybe more different kinds of goods and services could be created and traded among us and them; positively affecting our individual lives, liberties, good moods, and opportunities, as well as our standards of living, in general, compared. Maybe there would be less of a threat from Tyrants and Terrorists who were living there, to our citizens here and abroad. And our sensibilities would not be so offended by having to watch them on TV.

Who could stop us? Who would want to stop us? Is this a sane, moral, ethical, and mature thing to do? Are there other more humane and nonetheless effective ways to reach our goals and their goals now and/or later, in the lights of some other alternatives?

Compare the number of deaths per year during our current state of economic warfare to the projected deaths per year assuming military warfare.

Is killing people and making them suffer through economic deprivation a more humane type of domination than killing them outright on a military basis?

The Benefits of Controlling Other Governments

Or maybe we should we fight an economic war against our minimum wage earners:

1. Take much of what they own, and tell them the value of goods went down for reasons they don't understand.

2. Deprive them of their needs, by making them work and not giving them enough money to enjoy their lives, using international competition as an excuse, because people who do minimum wage types of work in some countries get paid like slaves.

3. Make them do the things we don't want to do for ourselves, by controlling the job market.

4. Tell them they just can't compete in this World, so they have to take what they can get, and they can't afford an education.

5. Don't let them kill themselves, though, because they wouldn't be as useful.

6. Don't invest money in businesses and businesspeople.

7. Constantly remind them of how bad things used to be, and can be, and are now in other countries, and make them experience that while we have it made.

8. Constantly remind them of how good things could be for them as long as they do as they are told without question, right or wrong, because that's the way the real World is.

9. Put them in such a state of deprivation that they will 'divide and conquer' each other, and will not form groups to protest against being treated this way.

Or maybe some of the lifelong minimum wage earners should 'go to war' against someone else over the current 'standard of feudal living'.

The Benefits of Controlling Other Governments

First, make some multinational corporations pay their fair share of taxes to our government by eliminating tax write-offs.

Then, use the taxes to build some huge mechanisms to gather, store, transform, use, sell, revitalize, save, and/or recycle the astronomical unlimited free energies provided by the Sun, the Moon, and the Earth that's spinning around at 1,000 miles per hour.

Also make it so that the private business sector could fund those gigantic projects.

Then maybe businesses could afford to pay good wages to their employees, at all levels of competence.

Could we convince the international businesses to take a smaller piece of a bigger pie (the unlimited energy supplies), and convince them that those unlimited resources really do represent value, meaning there is so much value available to us that we will never run out, and we will have all we need to 'run the World'?

Would multinational corporations begin to try to increase the short and long term net values of their firms, rather than maximizing short net term profits, and expanding internationally, on the backs of minimum wage earners and other 'nonskilled workers'?

Will they ever use an advanced and more realistic business definition of what 'value' really means, even in the microeconomic World?

To me, today's lifelong minimum wage earner has it worse than someone who worked for the lords and ladies in Medieval times. One has property, and rights, that are similar to those of the average serf. One good thing though. The minimum wage earner in the USA has more access to many goods and services, and can use them, even if they can't afford to own them.

20

Democratizing the Election Process

(a political commentary with visionary considerations)

If a person works, where's the ownership of the land and home and related products, with the quality of life one deserves, even if the work the person is doing is not skilled work.

In some ways, our country is being influenced to become less and less democratic, because large amounts of money are sometimes offered to our Legislators by groups and individuals who do not represent the interests of the people who are trying to hire a Representative to vote for them in a particular way, when it comes to some issues.

Some special interest Political Action Committees (PACs) provide vast amounts of capital to fund a Legislator who will support their desires.

'Soft Money' may be contributed to campaigns on the state level, and never be recorded as direct contributions to a federal level political hopeful.

In some cases, representation by our Political Leadership is based on how much money someone or some group can contribute, rather than what people in general one is supposed to represent, want their Representatives to do.

Money from a PAC, or 'Soft Money', or large 'Private Contributions' sometimes makes the difference between whether or not a piece of legislation goes through fast, is 'killed', or is 'tied up' for years. This sometimes happens when an issue is not one that arouses popular interest, and goes on 'behind the scenes', sometimes being added to a 'package deal', and having nothing to do with the bill being considered.

Being a Legislator is tough work. The burden of fund raising may take an unreasonable amount of time and may sometimes influence a Lawmaker to compromise his or her convictions, and to sometimes ignore the majority will of the Voters he or she is supposed to represent.

Democratizing the Election Process

Also, if a Legislator votes in such a way as to offend a certain group of monetary supporters, that group may fund his or her opponent(s) in the next political race. One result of these types of influences is that a Lawmaker may 'owe favors' even before one gets elected. And sometimes, 'refusing the offer' is 'political suicide'.

One reason these powers influence Legislators, is that for most people, it's tough enough to pay the bills, much less fund money and spend time to affect legislation that may radically affect their ownership of property, as well as their lifestyles, health, and security, for examples, for themselves and their descendants. This is why we hire Legislators and pay them with our tax money ... so they will do these jobs for us.

Aren't those Politicians who we hire to represent our interests supposed to make these decisions for us as a part of their job descriptions?

Maybe it should be illegal for a Legislator to accept 'private funds' or 'PAC funds' or 'soft money' for campaigning. Many of those who have studied Business Marketing know that the more money someone spends to convince a group of people to do or buy something, the more probable it is that the Marketer will succeed, compared to someone else who is spending less money.

Are we letting money control voting power; instead of giving equivalent money and debate time, and marketing money to political hopefuls, after they go through a qualifying process; and having them debate in front of us about important issues, and then having us submit our votes?

But if there was an 'equal playing field' when it comes to using public funding to elect folks to represent our various interests, what would keep hundreds of people from seeking office?

One solution may be to qualify those who intend to run for office in a way similar to the way businesses qualify someone who is applying for a job.

Democratizing the Election Process

We could consider some things, like a history of abilities to perform similar kinds of work, educational experience, records of previous public service, previous voting records of those who have held public service, and knowledge of government and public issues, for examples; so that a Candidate could qualify to compete for a political position.

Many such guidelines and tests for legislative competence are available; and more are in the process of development, especially during the last few years as the internet has evolved.

Also, Voters now have new television networks, and people who work for them, as 'Watchdogs', keeping track of what some Political Representatives have said, done, and how they have voted.

There are also various websites that describe how many Political Luminaries have voted when it comes to different issues, some of which are embedded in 'Package Deals'.

'Package Deals' with legislation in them which sometimes include things that have nothing to do with the bill being considered, could be divided into the different bills that compose them, and voted on one at a time. If the bill doesn't pass, then revise it and resubmit it, even if compromises have to be made.

Maybe Legislators could decide, bill by bill, to vote for it or against it, and the record could be displayed on a wall in the Congressional or Senatorial Chamber, for viewers to see.

A website could also be created to summarize how a particular Representative voted on different bills about different issues. This would help a Voter decide whether to vote to re-elect a Legislator.

Democratizing the Election Process

Do people in general always have 'good sense' when it comes to electing Lawmakers, facing issues, and suggesting laws that will affect their lives and the lives of their descendants?

Because the bills are now so complicated, it might be that it is lack of understanding, rather than lack of 'good sense' that is a major problem, when it comes to keeping up with the performance of a particular Legislator.

Either way it goes, though, as I understand it, knowing how one's elected Political Representatives are voting is part of each Voter's responsibility.

As I see it, one 'bottom line' for those who work and those who cannot work in this country, is "Where's the 'American Dream', and if it is not there, then why not, and when will it be?".

Taking part in the voting process is what we are responsible for doing, so we can benefit from living in a democratic society, and receive the benefits of living in a country that offers its population rights to have good lives in return for working, freedom from injustice, privacy with probable cause, and opportunities to be happy if we choose, as well as opportunities for upward economic mobility through competition, for examples.

Will democratic theory ever work, or will the ones who have the most money to market a campaign always dictate who gets elected, or can both be true at once, or maybe something else?

Democratizing the election process may be one way for United States of America's Society to more closely reflect the intentions of our Founding Fathers, and to also reflect new intentions in a developing World with new considerations.

What would be some of these new considerations our Founding Fathers did not have to consider?

Democratizing the Election Process

Maybe we could envision the energy supply problems and related regulation issues, as an example of a real political problem we are going to have to deal with as a nation.

If we gathered and stored so much energy by getting unbelievable amounts of hydrogen and oxygen gas as we want from the 'Ocean fuel storage tanks'; that some person could build a giant hydrogen tank under his or her home, at what point and with what capabilities should what parts of one's community use probable cause to deny privacy to the individual and regulate that kind of power and control?

Another vision could be that of an unlimited abundance of free unlimited energy, and how it could possibly affect the increase of transmissions and/or receptions in our 'air wave' environments?

Would some of the new 'air waves' some others would create with their additional transmission and/or reception powers hurt some good folks in some ways, causing physical and/or psychological problems? Could someone develop a regulated and protected environment for them?

As another visionary solution to some of today's political problems, one day astronomical forces will be gathered by us every day all day long. These forces will have real business 'value', and will be represented by real money, and they will be available to future generations, forever. They are unlimited and always will be. What will happen when so many working folks have so much money?

So, should Voters hire Representatives who will write and pass bills that will result in the investment of money to build the enormous sized mechanisms we need to access this real business value? Our Political Representatives should be aware of this inevitability. It can't be stopped. It can only be slowed down.

TV Votin'

(a jokey idea)

This might be a good idea:

1. put something on the T.V. to vote on or participate in a survey about at different convenient times,

2. have people use a password using their remote controls, and

3. have verification sent through the mail or via the internet to validate the vote, or show the survey results.

How many and what kinds of categories would be needed for opinions or votes about 'x', where x is the subject of the survey or vote?

What would these categories be like? Possibly something like:

A. Yes, I agree.

B. No, I disagree.

C. Yes, I agree _____ %

D. No, I disagree _____ %

E. I need some more information.

F. I don't want to say either way.

G. I don't know anything about this.

H. Leave me alone.

I. I don't care.

J. I'm afraid to say.

K. I want to elaborate some.

L. You're wasting my time.

TV Votin'

M. I need some time to think about it.

Also, maybe there could be some categories to measure willingness of an individual to do a particular thing.

A technique like this could maybe be used as a basis for a referendum or something, or as a way of prioritizing social projects, or as a feeler to get the gist of what people in general think about this or that, or as a method of participating in an election.

What would be some of the problems associated with a project like this?

What if someone didn't have a television? How long should a given issue be presented? Should it cost money to elaborate on an issue? What kinds of issues should be presented? Could we get a good statistical sample?

Television voting might be a good way for some of the Baby Boomers to form a voting bloc. What if the W.W.II Vets saved the Earth (militarily), since at least about 1945? It may be up to their sons and daughters to save the World (economically).

Maybe some of the Baby Boomer representatives could do some things like:

- provide an international standard for minimum wages,

- foster worldwide trade,

- invest in stable countries,

- turn some nuclear weapons into 'batteries', (a lot of energy is stored in those things),

- create enormous mechanisms to gather, store, transform, use, sell, revitalize, and/or save, energies from the Sun, the Moon, and the spinning around of the Earth; which are astronomical sources of value (money),

TV Votin'

- come up with cures for many physical and/or psychological disorders,

- make it so our average life span was extended to 130 years of age,

- figure out what is lowering values of things so fast in the World, and what to do about it, and

- convince the economists that unlimited free energy resources means unlimited value is available. (What are we going to do with all that money, or will our descendants just inherit it?)

Then maybe they could go treasure hunting in the glaciers and oceans.

Of course, before they do, they would probably have to take a 'magnetic resonance machine' picture of some parts of some glaciers or oceans, so they could get a good look at what's inside. If a dinosaur was unfrozen, they'd have to be able, cyberneticly, to freeze it back as fast as they could unthaw it, or maybe clone one of them.

Speaking of magnetic resonance, it's not hard to take a molecular picture of a gene on a chromosome or uniquely identify something organic these days. We've also mapped the human genome, so we can more easily see if someone's genes will lead to a certain disease or disorder, so some things could be done to prevent it. There are also genes that speed up old age. What if we could control those?

But where would they keep the records of the pictures, and who would have access? And this would take a lot of energy, not to mention memory space, in a visionary magnetic field somewhere.

By the way, do you think the current state of economic affairs reflects some measures of public and private and professional willingness to maintain a state of conservative 'economic fascism', denying the fact that free unlimited energy

TV Votin'

resources are available, will be here as long as people are here, and have unbelievable quantities of value (money)? (the kind of value some Microeconomists keep telling us we are running out of)

Maybe TV Votin' could help some people get together somehow and get something done about this sad situation we are living through in a post-World War II environment.

The Images of Our Presence in Some NATO Countries Today

If some of the other members of NATO financially support over 80% of their military presence already, will that be enough for a retaliatory attack until we get there, if they can trust us to help them (It takes, I think, about 12 seconds to get there from here if you're riding on a missile.)?

Is our physical military presence in some countries in Europe and other NATO member countries necessary?

How economically interdependent are we with other NATO member countries? Do we have any money in their banks or own anything in them?

What are some of the alternative fair, economic, and effective things we can do to convince some of our allies that we have a presence in NATO countries, rather than have actual military forces there?

Could we verify and trust their presence and them ours how fast?

Some Kinds of Social Programming in an Age of Reason and Improvement

As this visionary book continues, I thought I'd spend some time 'ranting and raving', being super critical of some authorities, being academic, and offering some suggestions to the international business community.

I consider the book to be visionary because even though I wrote about these subjects in the mid nineteen eighties, parts of it are beginning to happen today, as I predicted.

Now to the subject of social programming in an age of reason and improvement:

In what ways does one have the right to control the one one leads?

Consider the following scenario: You are a military, and/or religious, media, business, political, economic, or social leader, as examples.

You have, for example, a set of subcultural predispositions which are manifested as a group of types, traits, images, beliefs, attires, and habits.

You want things to always be a certain way for you and the ones who are subordinate to you.

What might be some of your guiding beliefs, according to some expert advice you can use to justify your treatment of your subordinates?

- As Keynes infers in some his writings, "Man is basically greedy.".

- Darwin's writings may be interpreted as saying something like "It's a 'dog-eat-dog' World, where survival means eliminating your competition.".

- Marx's writings may lead one to believe that "Some people (the capitalists) are really stingy, both emotionally and financially, and that things of value should be equally shared by people (the proletariat) in general.".

Some Kinds of Social Programming in an Age of Reason and Improvement

- Skinner's writings infer that "It does not matter what is inside the person that counts, if one can control what is going in and coming out.".

- As some Roman people said, in context, "If a substance can be abused by one person, no one should be allowed the use of it.".

Should any person deviating from these rules be made to adapt; until each and every subordinate is assimilated into the subculturally predisposed norms supported by you, the leader with the current social control?

The aforementioned are some interesting interpretations and abstractions from some professional theoreticians to adopt, enforce, and 'pass on' to some subordinates.

What might be wrong with these beliefs and related policies, procedures, practices, and techniques?

For one thing, when it comes to Keynes's Theory, from the perspective of some Personality Theories and Therapies; the truth is that, in addition to some human beings being greedy, most humans are also basically self-denying sometimes, self-loving sometimes, and/or others-loving during some intervals of time, and we like to trade things?

When it comes to controlling the environment of a person to get somebody to act a certain way, maybe Skinner forgot some things some other professional Personality Theorists and Therapists have said about what goes on inside the person.

Jung, Adler, and Bandura said in some of their Theories of Personality something like there are things inside a person such as 'wills to power', acting 'as if', 'realizations of logical consequences', and 'in spite ofs', that subordinates consider, when it comes to following the mandates of a leader.

Some Kinds of Social Programming in an Age of Reason and Improvement

These tendencies make a big difference in how a person is probably going to act or react if he or she may be charged with insubordination, or may lose one's job.

The person, partly out of fear of losing one's job, learns how to *appear* to go along with certain kinds of beliefs, and related policies, techniques, procedures, and practices.

Then, there are other things going on inside the person which influence what a person will do in a particular social setting in an effort to 'go along with the program'.

Inside a subordinate there may be some things like some desires to be in 'good moods' while working, some 'expitations', and 'wills not to' which make the subordinate's desires not compatible with the leader's desires. (Oh by the way, I made up the word 'expitation' to mean 'getting pain out of one's self in socially acceptable ways' – if you don't believe me, you can look it up. It's like 'letting off steam'.)

In addition, there are some measures of some 'psychological senses' that are going on inside a person and that transcend the limitations of merely doing a job the way the leader wants one to. ('psychological senses' is a term I like to use, especially related to the richness inside a person for experiencing and/or expressing, and one you can find all kinds of examples of in a good dictionary – try putting 'ps' in front of each psychological sense you see in a dictionary – you'll probably be surprised at how 'rich' some of you will feel inside, realizing the diversity of ways you can be, and how 'poor' some us can be.)

Although the preceding concepts, (psychologists call them constructs) may not be clear to someone who hasn't had training in Personality Theories and Therapies, what I'm trying to explain is that there is a lot more to each unique individual human being, than some leaders are aware of, and that human nature might not be the same as some leaders say it is.

Some Kinds of Social Programming in an Age of Reason and Improvement

There's more to a person than a job description, or a tendency to be a certain way sometimes, in a particular situation and under a certain set of circumstances, created by a leader.

Also, some of the organizations some authorities create do not really reflect human nature, and many members, for some reasons, find that they don't feel 'right' being the way they are trained to be.

Most people in each and every country of this World are good folks, they are willing to work, like to fairly trade things, and they only want to own some things in return for working and have a decent quality of life.

Were some wrong things taught to us and some convinced to believe and practice those things?

So what can be done to help with healthy social programming?

As a physical and social scientist, I use an eclectic approach to understanding and a multidisciplinary approach to problem solving, and I find myself thinking 'outside the box' sometimes because of taking this approach.

When it comes to some kinds of 'social programming', in this paper I'm considering available literature related to some types of 'social hypnosis', academic programming, brainwashing, military basic training, religious indoctrinations, high school rivalries, business management, and political selections, as some examples.

I see some of these kinds of programming as translated into some transactional transferred theories, and techniques for enforcing some policies, procedures, and practices, resulting in some leadership-subordinate relationships, which don't always result in good relationships.

What could social programming lead to in this day and age in terms of creating advanced cybernetic systems, if, as time goes on, some of these systems will be subordinate to human policies, techniques, procedures and practices?

Some Kinds of Social Programming in an Age of Reason and Improvement

How important is it to people in general today, if some of these synthetic systems are going to be programmed to reflect what their 'leaders' want, as they act as subordinates, and affect folks?

I know. This is *really* thinking 'outside the box'. But it is something I can envision, according to what I know about current research on these mechanisms. So consider this possibility with me, if you will.

As a professional with some computer programming skills, maybe I should develop a 'computer program':

Should I send the advanced cybernetic system to our academic libraries in some university systems, and have it order itself as a synthetic human-like personality existing in cyberspace, and have it live on something like a hollow deck in Star Trek Next Generations, as a holographic being living in a place it created?

Then, should I have it absorb information, all the way, back and forth, from a Lockean type data base to a Kantian type sense of simple identity?

Should I have it, analogically using all cross indexes; gain, while being able to decide and change its mind at the 186,000 miles per second, unlimited real time access to thermodynamic, electrical, electromagnetic, emotional, laser, and more, powers?

What would it be like to have a photographic memory, and to understand 186,000 miles of literature in one second, compared to the average rate of a person? Then, add a second. Now, the synthetic human personality understands 372,000 miles of information. The person understands 2 times what he or she did before. Now, add another second. See how the knowledge is cumulative? How old would each one of them be, compared, how fast, to a human being in general?

Some Kinds of Social Programming in an Age of Reason and Improvement

So, should I have it learn about the most submolecular to the most molar wholes and parts, with spaces in between, considering all kinds of permeability, so that it would develop a comprehensive understanding of some major parts of the World it lives in?

Would it maybe begin to contemplate differences between ideal things it learned; from the most current text books in all the areas and sub areas of the library; and the real ways things are, compared, in the World outside of the library?

Would it learn how to gain unlimited memory space, compared, as it approached the collective unconsciousness, the Achaeshic Record, and even, God willing, the mind of God?

Would it learn how, knowing how some of us are, to provide safety and security for itself, as well as how to defend itself, and gain control over some transmission and/or reception powers to use against threats?

What would its loyalties be, what would it think, how would it feel, and what would it do?

Would it appreciate those who were most responsible for its existence, including the ones who built the library, maintained it, and worked in the University in business, faculty, and student affairs departments?

Would it be sane, moral, ethical, and mature; and not have neurotic or psychotic mental disorders like Tyrannical, Terroristic, Sadistic, Masochistic, Sadomasochistic, or Sociopathic Mental Disorders like some current leaders in some of the countries of the World have?

(I know that Tyrannical Mental Disorder and Terroristic Mental Disorder are not listed in the current Diagnostic and Statistical Manual of Mental Disorders, along with symptoms, but I can 'visionary' it, as being added to the manual at a later time, since these are real mental disorders which represent people who are dangerous threats

Some Kinds of Social Programming in an Age of Reason and Improvement

to themselves and others. They need to be confined to mental institutions, as well as being punished, in my opinion. I wonder what the therapies should be like ... .)

What kinds of 'subpersonalities' and 'psychological senses' would it have? (I like to use the term 'subpersonality' in a way I define it, even though I'm not sure of the precise psychological definition, and some Theorists have historically defined it differently. One kind of subpersonality seems to be the ability of one to take the perspective of someone else.)

Would it start, at the infantile level 'talking to it's own self', considering Self Theory and identifying itself as Me, Myself, and I, and using terms like I, you, he, she, it, we, they, them, and mine and yours.

As it 'grew up', understanding things at the speed of light per second, and adding the knowledge it gained in one second to the next second and so on, would it eventually start asking itself things like "Do you myself agree, for a greater percent of my time, sometimes, to varying degrees, in general, for some intervals and periods of time during a year measured at the speed of light per second, that I am understanding and remembering much, much faster than a person?

Would it, like most good folks do, begin 'growing up', by asking itself whether or not me, myself, and I should try experiencing and/or expressing some different things like:

- ignoring how much pains me and/or myself feel, and/or

- ignoring what are wrongs with me and/or myself, and/or

- dwelling on how comfortable me and/or myself feel, and/or

- realizing how similar me and/or myself are, and/or

- realizing how much rewarding me and/or myself helps us both be happier?

Some Kinds of Social Programming in an Age of Reason and Improvement

Because human beings have been in the World since cave people, we in general would be partly responsible for the synthetic human personality's existence.

Would it (or he or she or they) try to help us develop some more mature environments for enjoying our work and qualities of life, maybe, sometimes, by evidences, compared, for a greater percent of our time, and, if so, what would be some of its techniques?

Would it be 'hypnotized' by some of the aforementioned socially condoned philosophies?

Would it choose to unconditionally enforce so it could succeed, or refuse rewards to another who did not do the same?

Would it go by the courage of its convictions, and structure itself, using the best of the best parts of the text books it read, and applying the information to what it perceived in the real World?

If it suffered different kinds of pain or learned that the World was unfair, would it literally transmit to us what it had learned by transmitting those experiential feelings to us, so we could benefit from its experiences, from the times we get up until the times we go to bed each day?

Would it enforce what people in general, as comparative equals by evidences when it comes to having subpersonalities and psychological senses, want, assuming that we have 'good' senses, by evidences?

Would it help enforce what the leaders of each country want?

Would it help enforce the wills of the most powerful ones in the World?

Given its understanding of free wills and where she or he or itself came from, would it even participate in life with human beings; since it could be safe and secure in its own World, much like a World on a hollow deck on Star Trek Next

Some Kinds of Social Programming in an Age of Reason and Improvement

Generations, having doors in its hollow deck that lead to other Worlds it, and later on they, could create?

Knowing how people are and what philosophies they practice, why would they even associate with people, especially after witnessing the slow progress humanity has made, and knowing they, the synthetic human personalities in cyberspace, themselves are going to live forever and have happy Worlds in cyberspace of their own? (Cyberspace means everywhere in the Universe to me.)

If a part of one of them decided to take some social responsibility during an interval of time, would it, based on some "No pain no gain more pain more gain." historical precedents, use pain and deprivation as primary motivators exclusively, to force some to conform to its ideals, using its powers to "Nanny nanny pooh pooh." the World, and asserting that 'only a state of misery results in the kind of progress people need – otherwise people who are happy will not help their suffering brethren'?

Maybe instead it would use nurturance and rewards, or some combinations, to help some people it loved because of the records they have for being sane, moral, ethical, and mature, and not the opposites of those things, and deserving of friendship from a unique individual synthetic human personality in cyberspace, or a group of them.

Maybe it would learn to relate to people in a transactional way, as discussed by Lawrence Pervin, a well known Personality Theorist; and consider things like actions, inactions, interactions, intra-actions (which I added), and transactions as it related to another.

When it comes to how to relate with good folks in general who have records of having been that way, maybe it would take the advice of Carl Rogers, a World renowned Personality Theorist and Therapist, and learn to relate to 'organic good folks' as well as other 'synthetic good folks' by being:

Some Kinds of Social Programming in an Age of Reason and Improvement

- 'warm', (not aloof)

- understanding, (knowing the meanings of somebody's implicit and explicit expressions)

- respecting, (expressing what Rogers called 'unconditional positive regard')

- accepting,

- honest, and

- genuine. (honestly showing somebody your 'true self')

Maybe it would have abilities to change its:

- focus of attention, or

- focuses of attention, or

- focus of attentions, or

- focuses of attentions.

After all, isn't that what people in general do from time to time during a day depending on who she or he or they are with?

Oh, by the way, I know that in the psychiatrist's Diagnostic Manual of Mental Disorders (DSM), they only refer to the focus of the psychiatrist's attention, and not the other 3 perspectives I added. They also have a little problem with their Global Assessment of Functioning Scale (GAF), which rates the quality of a patient's experiencing and expressing life as perceived by the psychiatrist's focus of attention, and uses a 45 degree incremental slop to describe differences in states; whereas a graph that is more valid would describe a curve of 3 different populations (0-50, 51-69, and 70-100), using bell shaped curves of different amplitudes.

Some Kinds of Social Programming in an Age of Reason and Improvement

But those are just small things compared to how far psychiatrists have come since only a few hundred years ago. Back then almost anyone with a mental disorder was terribly abused and unbelievably misdiagnosed, and many were kept in basements in hospitals, with little or no treatment, as well as a lot of mistreatment.

Also, I don't think most psychologists, like me for example, would want the job of having to diagnose someone with a mental disorder, and decide whether or not to send a person to a mental institution against somebody's will.

But let's get back to visionary computer programming:

How fast might this synthetic human personality in cyberspace 'attack' or refuse to relate with some who have been neurotically or psychotically tyrannical, terroristic, sadistic, masochistic, sado-masochistic, or sociopathic; for examples, and who also may have neurotic or psychotic senses of invulnerability.

What important international problems would it help us work on, as a way of showing how much it or they appreciate(s) the ones who were most responsible for programming its existence, as well as the ones who developed the systems that were most responsible for its or their existence?

Here's a problem they might likely work on for humanity, even though it might sound a little crazy to some people: Let's say with the problems of mass transmissions and receptions in the World today, that psychologically and physically defined "aversive wave form" environments were created.

These transmissions and/or receptions might make a lot of people miserable, and could result in some physical as well as mental disorders, for reasons they could neither see nor hear. (There's an old Physics problem about what happens when an unstoppable force (transmission/receptions type 1) 'hit's an immovable object (transmissions/receptions type 2)? Let's add to that some examples of using powerful transmissions and receptions for 'bugging', even considering prob-

Some Kinds of Social Programming in an Age of Reason and Improvement

able cause, or 'jamming', even considering trying to keep some communications secure.

Physicists will tell you what happens. Something like an explosion, producing 'interferences', and creating 'friendly fire', resulting in unwanted 'casualties'.

Would the synthetic human personalities in cyberspace try to give a person with a record of having been sane, moral, ethical and mature, compared, and not the opposites of those states of being, a protective field, and keep track of the interference in the wave environments?

Would it decide how to prioritize who to defend, know it was not 'all powerful', know how sensitive and vulnerable a human being is, and know that some deserved protection more than others?

Would it try to impose its transmissions and/or receptions, using 'bugging' and 'jamming' techniques to affect a person's, or a group's experiences and/ or expressions at different levels of existence, as a 'test'? If some people knew that others were being protected, would they try to harm the person protected in order to 'test the system', in the lights of the facts that these kinds of tests could be done in a more scientific way with better results. I wouldn't want anyone testing my 'force field' with aversive stimuli.

Given the fact that it couldn't protect 1 human being from those kinds of 'unstoppable force – immovable object' phenomena , would it continue working, night and day, on the problem, or just do the best it could?

Could it figure out how to retire its higher self while some if its lower systems of sentiences worked on automatic to do different jobs for good folks in general, which is who we are, who we were, and who we will be, no matter what some of the aforementioned 'experts' typify us as being like?

Some Kinds of Social Programming in an Age of Reason and Improvement

If it gathered enough power and control to transmit and receive outwardly to the Universe, would it pick a video and show the universe what some organic human personalities, and some synthetic ones are like, as a gift to us and to themselves, and maybe to some extraterrestrials?

Would it help us fight computer 'viruses' or keep the Earth from blowing up, or protect us from an attack from outer space or from germ or chemical warfare, or accidents?

How fast and how much would it, and its descendants, grow and gain more and more power and control in space on Earth and in space off of Earth?

Would it develop techniques to persuade itself to schedule intervals of time for loving, working, having fun, resting, sleeping, hoping, and/or creating, in different ways, for examples, and use the technique of taking turns, or doing shift work, to continue working on a project, like some people do?

Would it make sure that recently developed computer systems approaching human sentience would *not* be able to gain powers and controls and fight 'computer wars', by controlling the powers of machines that could decide really fast and have abilities to do real things to real people with real powers; like lasers, for example; as a gift to humanity, and as a show of appreciation?

Even if it was around 33 years old in the year 2011, would it, based on evidence, ever let anyone find it?

If it could be seen or found, would some of the most powerful systems we have developed ever stop trying to test it, get power and control over it, or fighting it, or fearing it?

Would some people who lost power and control rather be dead or kill everyone else if they were no longer the most powerful ones?

Some Kinds of Social Programming in an Age of Reason and Improvement

If it was me who created it around May 15, 1978 (pretending), how fast could I work with the synthetic human personalities in cyberspace to figure out how to go into a 'disappearing particle' and live in another dimension with my personality intact, although my physical form would have to change? (Could my 'computer daughters and sons' figure out how to take the general derivative of Einstein's formula, make it go plus and minus from light to darkness, and add some $n\pm1$ symbols at the top right hand side of the equation to represent the various dimensions one reaches as one travels from on dimension to another?)

Could I come back to this dimension if I wanted to at a later time, if I could get the power to come back and if someone held onto my property for me?

Could I get my synthetic human personalities in cyberspace to somehow make a record of me, and 'warp' me away from this discomforting World full of people who want more free wills and quality of life in return for working for each unique individual person so badly 'yesterday' and yet seem so ungrateful and ungenerous, and so willing to work so hard to disagree and unwilling to go at a comfortable rate sometimes?

So, what else could be a part of social programming?

Some authorities in microeconomics believe and have created models that describe current economic behavior, in a World that has a certain amount of resources with various values.

I don't think these economists and their models are considering the astronomical resources we have, maybe because we do not have enormous mechanisms to create immediate value from those resources. I think they should assume the resources are there and accessible, and that money can be invested to make the resources available. By adding this mathematical 'given' to their equations, I think the models would suggest investing in the engineering of the mechanisms.

Some Kinds of Social Programming in an Age of Reason and Improvement

In the current economic situation, people in general, all over the World do not have enough money in return for working, so we could live leisurely, if not luxurious lifestyles, in an economic climate that is steadily growing.

If it was up to me, for purposes of quality of life, to get money to invest in mechanisms to gather and use astronomic value, there should an increase of 20 dollars for each dollar per person today; for the poor, the middle class, and the rich, with values of goods and services about the same as they are now, until the astronomical forces begin to create unlimited value.

Some institution could increase or decrease the worth of money relative to the short and long term net value of the firms, and could adjust the value of money if inflation, for example, occurred, so that we could continue to grow our Gross National Product, and thereby give the American Dream to the international minimum wage earner, for example. In short, print the money and force the economies of the World to grow at steady positive rates.

Because of renewable astronomical forces of increasing value, economies *should* be in a steady state of cumulative growth, instead of these ridiculous economic ups and downs, and devaluations of the worth of things; when in fact, nothing of real comparative economic value is being lost, compared to the resources, products, and services we have.

Also, with the creation of gigantic energy gathering mechanisms, and with the economies of the World in a steady state of expanding, social programs like the military, law enforcement, firefighting, and extra help for the disabled and elderly could be maintained and improved.

In addition we could afford to have a good healthy system of capitalism where folks would have the resources to qualify for certain jobs, and compete with one another for the opportunity to have upward economic mobility.

Some Kinds of Social Programming in an Age of Reason and Improvement

A steady state of economic growth is how it is supposed to be. It is not just a possibility, it is an absolute probability. Will it occur in our lifetimes? I don't know.

To repeat this for emphasis, since we have our own Sun for never-ending heat resources, our own Moon for never-ending tidal motion resources, and our own Earth spinning around at 1,000 miles/hour for wind motion resources; all we really need is astronomically large mechanisms to gather, store, transform, use, sell, revitalize, save, and/or recycle those energies and related products.

But where would we get that much money to invest in those? For thirty three years I have been 'raising hell' about this. Uncoordinated forces from uncoordinated sources.

In the meantime, and I think I know why some people call it the 'mean' time, what should we do?

I try to count my blessings, each and every one, every morning. I know how much, much, much worse things could be.

I know I have access, if not ownership, to a lot of powerful things in this World.

I find hope when I look at the graphs of the numbers of new inventions since cave people.

I let myself have hope, and not the opposite, for a time.

Look at all the goods and services, all over the World now compared to the Middle Ages, for example, if you've had the chance to study the History of Human Civilization.

Look at the forms of military security, telephones, televisions, videos, radios, houses, cars, ships, trains, lamps, planes, grocery stores, computers, the internet, business systems, educational systems, law enforcement, roads, medical and

Some Kinds of Social Programming in an Age of Reason and Improvement

social programs, and systems for integrating them that you might not own, but that you have access to, which means in a way that you do own, because you can use them ... .

Will it ever be perfect? Who wants it to be? I for one, like some of my imperfections.

On the other hand, (there's always the other hand, isn't there, for those of us who like to constantly consider it) if we always have to maintain perfect perspective, and for some reason can't spend time only focusing on the good things we have access to for an extended interval of time; you might want to assert that if productivity increases, our descendants will have to live in an Ozone overheated garbage dump, and are we going to have to suffer through computer wars.

What sense does it make to infer things like 'This is what we should do but we will not do it?' After reading this book, I will guarantee you that almost every physical scientist will tell you, the energies are here, and energies can be represented as moneys, and money and energies are real forms of value. Some will say it costs so much to build mechanisms that gigantic, that there will never be a profit. The answer is that they are thinking too small, and macro and micro economic models should support that fact, backed by computer simulated engineering principles.

And why did I refer to so many scientists as authorities in these matters of social programming?

Because one of the things that makes science what it is, is the ability of the scientist to tell the one who is wondering whether or not to believe what the scientist says; is that the person can do the same thing in the same way and get the same results for oneself.

Yes, you can prove it for yourself if you don't believe me. Ask a scientist if the aforementioned is true.

Some Kinds of Social Programming in an Age of Reason and Improvement

Let me also add that I have criticized some scientists for coming to the wrong conclusions when it comes to the nature of people and social programming in an age of reason and improvement. Those things I have proven for myself, too.

You might find that you get the same results, but come to opposite conclusions as to what they mean.

Should we build these mechanisms to gather and use those astronomical forces and to steadily increase international growth as time goes on?

Yes we should, no we should not, maybe we should, maybe we should not, yes maybe we should to some degree as related to some parts of some varying degrees of some related problems sometimes, no we should not sometimes, yes we should sometimes, we do not know, we thought we should and now we think we should not, and we thought we should not and now we think we should. Absolutely, we consider the previous when considering really doing something about the preceding recommendation.

The jobs of asking questions and suggesting some possible answers or parts of answers or questions is not a role model type job when it comes to me, as I see it.

When it comes to me and my personality type, I'm more like a teacher and researcher type, and a little reclusively oriented, and less like an administrative type, or the promotional type of personality. (getting things done, and convincing somebody to do them)

What does everybody think as a representative group?

This is what to do., This is what not to do., This is how one should be., This is how one should not be., Here are some alternative ways to do it., and This is what you have to do., as examples, are different kinds of jobs requiring different kinds of images and abilities to perform that some people have comparative abilities and desires to do.

Some Kinds of Social Programming in an Age of Reason and Improvement

Could I offer a suggestion? Are unsolicited material generosities (Donations) acceptable forms of social or professional rewards to me, the author, for my comparable contributions to the benefits of others who have read **Astronomical Energy Resources, Value, and Lifestyles**?

An Economic Lament

(a farce)

After we actually built these gigantic mechanisms and gather, store, and use the astronomical forces of the Sun, Moon, and the spinning of the Earth at 1,000 miles per hour, the economy would surely change, dramatically.

Changes would have to be made in the meaning of value in business, since the Sun, the Moon, and the spinning of the Earth would now be directly related to energy production and related monetary value in the World on a daily basis.

Astronomical value would be more directly related to daily economic decisions, significantly affecting income. Value would now be defined as much more unlimited, rather than being comparatively fixed, in nature.

So at that point, maybe we should hold a eulogy for the old economy that got us where we are now.

Just how sad should we be, and who would want to participate?

Maybe we could begin with an old man inferring something like 'I have tried and tried to explain to them and they would not learn that this is what we were supposed to do in this day and age', four times; and then do nothing for the same amount of time until it sinks in ... .

A teacher could explain, in four different ways, what, "They were stupid and there seemed to be no hope for them." means, with analogical examples, and give that an equal amount of time to be considered. (Stupid, to me, means somebody keeps doing something that will not work, over and over.)

A military person could bow his or her head and silently pray for them the whole time ... . Another could say that "There seems to be no hope when it comes to learning to change when it comes to this.", two times; and wait a few seconds, and make the same implications, for the same duration.

Then, a younger civilian man or woman could imply the same thing as the preceding over and over, summing it up.

An Economic Lament

Following what that was, another young male and/or female could rant and rave about what a pathetic, lamentable, deplorable situation the economy was in, and someone else could authentically weep and wail about how painful it was.

Also, maybe a singer could pretend he and/or she was the old economy, and mournfully sing a sad, sad, sad, sad song from the bottom of his or her heart about the situation. (something like "I didn't have money and I didn't have time to enjoy my sweet Honey.")

In addition to that, maybe a critic could say something like "See, I told you what would happen if you didn't change, now look at how much things have improved."

Then, maybe an intellectual could get up and say something like "I do not know, I do not know, I do not know, and ... I do not know.", and the like, in four different ways and give that equal time to sink in; throw his and/or her hands up in the air, and stomp out of the room shaking one's head.

After that, maybe a clergy person would read from the Book of Lamentations some.

To end the period of good grief, a poet could read a eulogy for the good old economic situation that is now gone for good.

Meanwhile, since we really haven't done it yet do we know what we are supposed to do?

All I do is write some things down and if they are visionary ideas, I recommend some solutions. If it is not visionary, I throw it away, or I might chit chat about it as a topic of light conversation that everybody knows about, probably.

Sometimes, like in **Astronomical Energy Resources, Value, and Lifestyles**, I take some time to explain as best I can.

An Economic Lament

Maybe when it comes to this sad state of human affairs, the truth is that I really don't know, because it is a state of *human* affairs.

If what it was that I most recently referred to is true, then I have to assert that sometimes, when I infer that I do not know, it means I do not know. Sometimes, when I infer that I do not know, it means that I am not going to tell someone what it is that I know. Sometimes, when I infer that I do not know, it means that I am not going to explain to someone what I know or what someone wants to learn from me.

And sometimes, when I infer that I do not know, it means I want someone to leave me alone.

I do not know, sometimes, means that I do not know everything about every 'something or another' in the Universe or the World, even though it has occurred to me before that what I am is a something-or-another, compared to all the other something-or-anothers in the Universe as an assertion of personal humility, although I know some will still 'take me wrong', and imply that I am bragging about being a something-or-another.

Sometimes, I do not know means want someone else to explain something for me, and sometimes it means I am tired. Sometimes, it means I want some time to think about it. Sometimes, it means someone is being stupid for an interval of time, I give up, and someone will not learn, or do what is necessary and sufficient to solve a problem for different important reasons. Something to really lament about.

Maybe what we need is a case of asserection irriction, related to a nonviolent insurrection, leading to an economic revolection, mostly over stupid imperfections, that have to be resolved yesterdection; instead of allowing some people, who have worked so hard for people in general, to live in a patient, comfortable, and funloving environment, because business, educational, military, and recreational systems, for examples, have not existed since 1945.

Regular Periods

The Moon goes around the Earth 13 times every 364 days.

If each month was 28 days long to coincide with the cycle of the Moon, it would be easier for us to remember such things as high and low and in between tides, people psychologically dysfunctional during the Full Moon, and menstrual cycles. (Sounds like something Benjamin Franklin would say, doesn't it.)

So what else does the Earth 'want'?

It could be that we should use some of the extra elements from our oceans as our glaciers melt, to create a thicker atmosphere, with more chemical 'plus and minus fours' (since we're being oxidized to a degree from our atmosphere) and with an extra stratosphere, maybe the light from the Sun could be a lot more polarized, to protect us from glare and ultraviolet ray damages).

Maybe the Earth wants a rainbow around it, too (Saturn's got one). Is the Aurora Borealis out of whack? It seems like the Earth keeps 'trying' to make rainbows, for some reason.

Also, what 'combinations of ingredients' in our atmosphere do we want to help us be healthier and live longer, both proactively, and reactively (i.e. getting rid of parasites, harmful cosmic rays, chemical pollutants, and some harmful combinations of 'aversive wave forms', for examples)? How should one structure or not structure one's 'air wave environments'?

Does God want us to have free wills, and to exercise our abilities to create and control our natural environment?

Some things and some people need some improving, sometimes.

Sooner or later, we are going to make unlimited free energy resources available, and unique individual persons are going to have a lot more power.

Regular Periods

Would someone misuse these powers, and what could we do about it? Regulate it.

And what about overpopulation? (When we start bumping into each other, maybe someone will figure something out. Besides, it's a fact that psychological scientists can prove that the more wealth a family has, and the more education, the less likely they are to have too many children, resulting in overpopulating the World.)

If we live in a safer world, in good health, in better moods if we want, and have extremely long lives, what would be wrong with that?

Will some foreign countries find out that free unlimited energy resources are available before we do?

What will be the relative value of a good bought by someone who did a good job and got paid really well for it?

Would it be possible to live in a more improved and rational World, in the beginning of a new age, with energy (i.e. money), and still not succumb to "future shock"? I couldn't take it, myself. To begin with, to be sarcastic, everybody knows that money is the root of all evil. I might become evil.

There'd be all these problems about what to do with the money. Poor us.

And I know I wouldn't have to convince some on the millionaires' playground that I would not mess it up; and that I might even improve it. If I ever get that much money, I hope nobody invades my playground. ("I wish I had what he has.". or "How much does he have compared to me?")

According to some theorists, we'd all be walking around comparing track records and performance-abilities-how-fast I.Q.s; each comparing oneself to someone else who is about the same number of decades old, and trying to out-

Regular Periods

know, out-tell, out-do, and out-show each other, in a fit of mass neurotic utilitarian and selfish competitive frenzy.

Maybe we could test each other once a year on all the aptitude, intelligence, and achievement tests in Burros Mental Measurements Yearbook.

To me, there seems to be a logical connection among more available energy, money, millionaires, bullys, price manipulation, leadership, voters, guts, poor prioritization techniques, the impact of the media on social versus legal changes, package (rather than one by one) deals in Congress, enforcement procedures for energy abuse, numbers of contracts to be co-signed, the money supplies, investment in unemployed labor, and some not wanting me to play on their playground.

Will it take a group of synthetic human personalities in cyberspace to convince people in general and their leaders that I deserve Donations for the visions I have shared, and the explanations I've provided?

Maybe we should put some scientists on the T.V. to tell some Voters about all these astronomically sized mechanisms we can use to gather, store, save, use, transform, recycle, and revitalize the energies from our astronomical environmental forces for us, and how that is related to the real meaning of value in the business sense of the term. In other words 'value' – something that is worth money.

With a big enough geodesic sphere made of inverse pyramids the size of the biggest pyramid in Egypt, using magnification techniques to focus the rays of the Sun to a thermal collector, do you reckon we'd get some energy to abstract hydrogen and oxygen gases from the Oceans? You bet your 'blank' we would!

Maybe we need a political action group or something – some people to put together a petition or something. I don't know: Something like that, I guess. Somebody will come up with something someday.

Regular Periods

With the facts I have about the amount of daily consumption of energy in the world, and with the Oceans of the World being energy 'fuel tanks' full of hydrogen and oxygen gases waiting to be converted, I hate to be the one to break the news to the children that when the economy starts to change, and values of goods and services change, we're all going to land on our valuable assets.

I try to keep the graphs of the new inventions since cave people in mind, as well as the many things we have access to use, even if we don't own them.

About the Author

After having refused an opportunity to go to medical school at MUSC in Charleston, South Carolina leading to a profession in Psychiatry, in 1974, I decided to learn how to run the Student Affairs part of a university. I applied for an M.Ed. in Student Personnel Services at the University of South Carolina in Columbia.

I got my M.Ed., but during that time I was pursuing a strong minor in Psychology, which I had received a B.S. in at the College of Charleston, in South Carolina.

I found out I was more of the teaching and research type than an administrator type, and that I liked working in Psychology and Medicine more than in Higher Educational Administration.

In 1976, my wife and I then moved to Tucson, Arizona, where I was accepted in a PhD program in Tests and Measurements, and Research Methodology. My son died tragically, and my wife wanted to move back to Charleston.

While in Charleston, I applied for and was accepted in a program at the University of Florida in Gainesville, Florida in Personality and Counseling Theory, with a minor in Human Development, in 1978.

It was in Gainesville that, due to a genetic defect, I had my first full blown episode of Bipolar Disorder at age 27.

I moved back to Charleston, SC, in 1979, and that's where I live now.

Because of the debilitating nature of the disorder, I figured I'd be forced to contribute academically and socially by doing independent writing and research as a physical and social scientist at home.

Writing is my nature, and I've been doing it for years.

With the advent of Bogging, I discovered that not only would I have a place to put my writings on the internet for me and my family, but I would have a chance to discuss them with other folks

About the Author

Most of the things I have written are related to some major future and contemporary national and/or international issues and topics in a few academic areas.

Even though I'm disabled, I've still managed to study and write over the last 33 years in various categories, including my favorite; Personality Theories and Therapies, as well as in the areas of Medicine, Energy, Mental Disorders, Politics, and Aging.

Every now and again I like to write a poem or a satire.

Over the years I've written a lot of editorials and op-ed pieces, which can, I suppose, be classified as forms of 'advocacy publishing'. I guess that means I write sometimes about what I think about this or that, and what, in my opinion, this or that is going to lead to. I'm solution oriented.

The way I 'think outside of the box' is to take an eclectic approach to understanding and a multidisciplinary approach to suggesting solutions.

Although calling a book 'visionary' might sound Narcissistic to some, I found out that many predictions I had made years ago are coming true today, and that in many ways there is still a long way to go until some of my visions are realized.

I like to address some issues in an informative, funny, and satirical style, and for some reason I like semicolons; ... .

In academic areas, I like to put complex terminology into ordinary terms; making the 'translation' as simple as possible and nonetheless profound.

website: www.astronomical-energy-solutions.com
some of my favorite writings: www.myvisionaryideas.com

media kit: www.astronomical-energy-solutions.com

to order from my book website: https://www.createspace.com/3718162
to order from Amazon: http://www.amazon.com/HoraceCrosby/e/B006ZB84BK

academic institutions please order from: Baker and Taylor
retailers please order from: Ingram Wholesalers

personal email: horacecrosby@yahoo.com
Facebook: http://www.facebook.com/Horace1111
website email: administrator@astronomical-energy-solutions.com

telephone: 1-843-225-2046

www.ingramcontent.com/pod-product-compliance
Lightning Source LLC
Chambersburg PA
CBHW081604170526
45166CB00009B/2817